Milady's Razor Cutting

Kenneth Young

Milady Publishing Company
(A Division of Delmar Publishers Inc.)
3 Columbia Circle, Box 12519
Albany, New York 12212-2519

PUBLISHER: Catherine Frangie

ILLUSTRATOR: Kevin Buchan

PRODUCTION: John Mickelbank

COVER DESIGN: Susan Mathews

For information address:
Milady Publishing Company
(A Division of Delmar Publishers Inc.)
3 Columbia Circle, Box 12519
Albany, New York 12212-2519

Printed in the United States of America
Published simultaneously in Canada
by Nelson Canada
a division of The Thomson Corporation

1 2 3 4 5 6 7 8 9 10 XXX 00 99 98 97 96 95 94

ISBN 1-56253-180-8

Library of Congress Cataloging-in-Publication Data
Young, Kenneth, 1952-
    Milady's razor cutting / Kenneth Young.
            p. cm.
    ISBN 1-56253-180-8
    1. Haircutting. 2. Hairstyles.
    I. Milady Publishing Company. II. Title. III. Title:
    Razor cutting.
    TT970.Y69 1994
    646.7'242--dc20   93-2854
                                          CIP

Directions for Use of This Book

This working manual was created in order to give you real life practice in cutting hair with a razor. The technicals in this book are actual renditions of salon cuts that the author has created for his clients.

Each technical is designed to give you practice working on a specific type of cut, and different types of hair. *Milady's Razor Cutting* includes men's and women's ranging from the conservative to the trendy. There is also an implements key that will help you determine which styling implements you will need to use.

The last page of the book contains a scoring guide for your instructor's evaluation of your performance of these techniques.

Keep practicing each until you are comfortable enough to add your own unique elements to the finished cut and style. After you have mastered these techniques you will be well prepared for the haircutting challenges you will encounter in your professional salon life. Enjoy and remember—never stop learning.

IMPLEMENTS KEY

Razor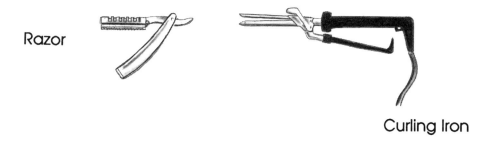

Curling Iron

Large Diameter Round Brush

Small Diameter Round Brush

Roller

Blowdryer

Tail Comb

Vent Brush

Velcro Roller

Pic comb (This is a combination comb and pic.)

Cutting

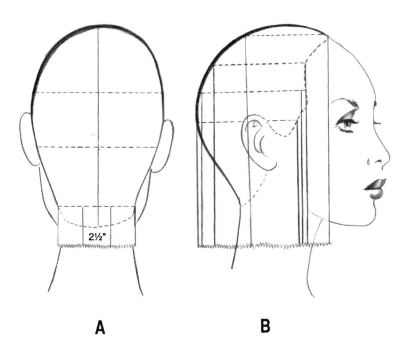

A

B

A Section the hair into four sections from the center of the front to center of the nape. Divide the head over the top of the head from ear to ear. This will give you four sections. Create a triangle bang section from the center of the eye to 2 inches back from the hairline. Keep the head in an upright position to cut the back. Use a horizontal parting to establish the guide line in the nape. Start at the center of the nape and cut the hair to a length of 2 1/2 inches. Cut the hair parallel to the floor working out from the center of the nape. Continue using horizontal partings and 0 degree elevation to complete the two back sections.

NOTE: When cutting a one length cut with the razor, short strokes should be used.

B Create a guideline on the left side using a horizontal parting at the bottom of the side section. Match the length of the side section and cut. Make sure the head is still in an upright position. Continue working up the head using horizontal parting and matching the length.

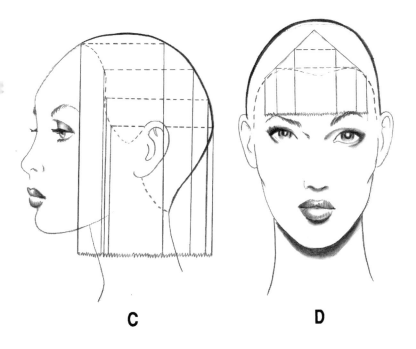

C

D

C Cut the bangs slightly longer than the eyebrows. Use horizontal partings to subdivide the bang section if the hair is thick. Cut the bangs straight across. Then lightly texturize the bang section by laying the razor flat to the strand and using light pressure to remove varied lengths.

D Taper the last 3 inches of the hair on the sides and back by placing the razor on the hair shaft at a 30 degree angle. Then, using light pressure draw the razor to the end of the strand. This will weaken the outer portion of the hair to help encourage it to bend away from the head. If the hair is very thick, you might have to subsection the head to get the hair to respond. Start at the top of the head and work down using horizontal partings. If the hair is a fine texture, the top section will be enough.

3

Styling

E   Dry all the hair with the exception of the last 3 inches from the ends. Direct the hair up from the head to create lift and volume throughout the head. Do not dry the bang section at this time.

F   Start styling the hair in the back. Divide the back into two equal sections from the top of the crown to the nape. Use a round brush 2 inches in diameter to begin the flip. Use 2-inch partings and roll the hair one turn around the brush. The hair should be rolled up to encourage the flip. Work up the back alternating sections.

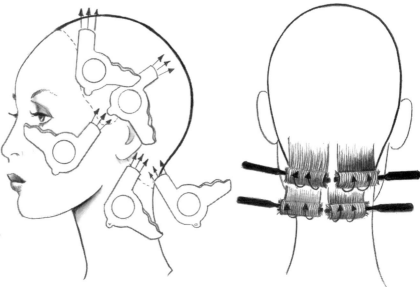

E

F

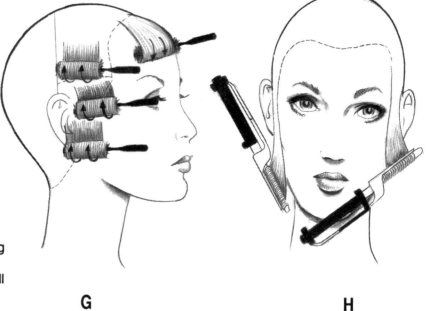

G   Repeat the same procedure on the side sections. When drying the hair ends, be sure to direct the heat at the top and the bottom of the brush. Allow the brush to cool then remove the brush. Dry the bang section using a vent brush. Do not elevate the bangs when drying as they will lay close to the head.

H   Using a 3/4-inch curling iron, wrap the hair on the sides and in the back one turn around the curling iron. Do not subsection horizontally at the sides and back as this will create too much volume at the ends.

G

H

Cutting

A Section the hair into four sections from the center of the front hairline to the center of the nape and over the top of the head from ear to ear. Start in the nape. Use diagonal partings approximately 45 degrees angled down from the center of the nape to the hairline. Push the head forward to create an undercut. Using short strokes with the razor, cut the hair at 0 degrees. Follow the same angle as the partings to create a short length at the center of the back (approximately 2 inches) to a long length at the sides of the back. Alternate cutting the partings from left section to the right section while cutting the back. Compare each side after cutting a parting to make sure the length and cutting angles are consistent.

B Use the same partings and cutting angles to cut the right side. Make sure the head is turned to the side and pushed back to create an under cut. Match the length on the side to the longest length in the back and cut. Remember to use short strokes with the razor.

C Repeat the same procedure on the left side. After cutting the first parting compare the cutting angle to the right side. Then compare the length to the first parting on the right side. If both match, the cut will be even.

D Create a horseshoe bang section from the corner of each eye and 2 inches deep. Using horizontal partings in the bang section, cut the hair to the bridge of the nose. Once again, use short strokes with the razor to create a tapered but strong line.

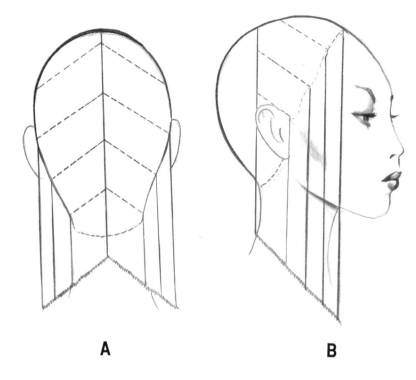

A

B

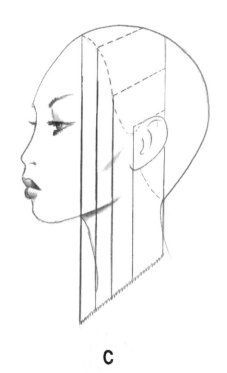

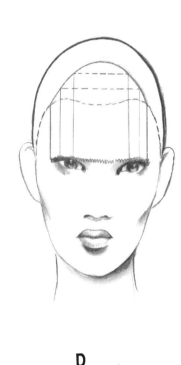

C

D

Styling

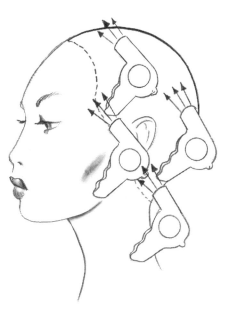

E

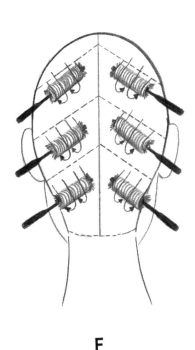

F

E   Remove most of the moisture from the hair by directing the hair up and forward. This will create volume and allow the hair to fall forward accenting the angle. Leave the ends of the strand damp.

F   Divide the back into two sections from the top of the crown to the center of the nape. Use a round brush 1 1/2 to 2 inches in diameter to dry the ends of the hair. Wrap the hair around the brush at the same 45 degree angle as the cut. Subsection the hair into three equal parts on each side of the back. Dry one subsection at a time starting at the bottom and working up to the top.

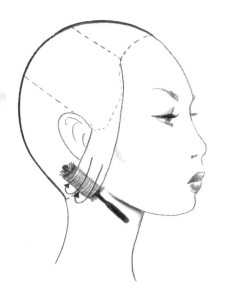

G

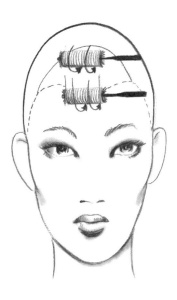

H

G   Using the same angle as the cut, subdivide the side sections at the same angle as the cut. Once again, roll the hair around the brush at the same angle as the cut and partings. Be sure to dry the hair from both the top and bottom of the brush. Allow to cool and remove the brush. Dry the bottom subsection first and then the top.

NOTE: If the hair is exceptionally fine, subsections when styling might not be needed. If the hair is exceptionally thick, more subsections will be needed.

H   Use a 1-inch diameter round brush to dry the bangs. Divide the bangs into two sections horizontally. Dry the section at the hairline 1/2 off base and under. Dry the second section on base to create volume.

Cutting

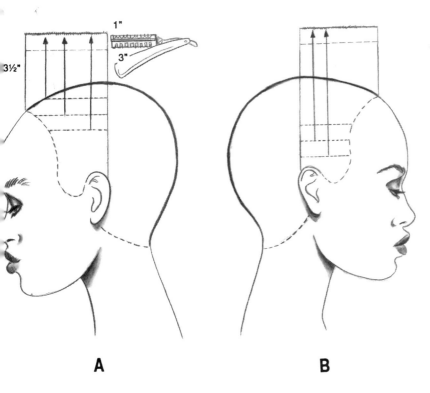

A

B

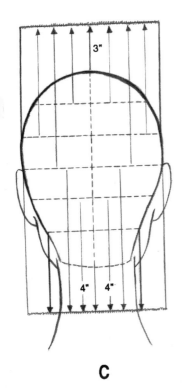

C

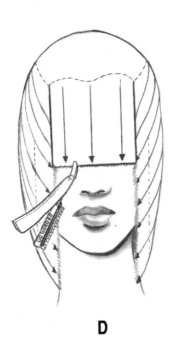

D

A This cutting is to be completed on chemically relaxed hair. Section the hair into five sections. Divide the head over the top from ear to ear. Divide the back from the center of the top of crown to the center of the nape. Divide the front into three sections, by parting from the center of each eye straight back to the top of the crown. Make a 1-inch parting from the top of the crown to the front hairline in the center of the top section. Cut to a length of 3 inches at the back of the section to 3 1/2 inches at the hairline. Continue to use vertical partings in the top section (from front to back). Pull the hair up to the guide and cut. Use a 1-inch stroke on the razor to create a lot of texture at the ends. To avoid cutting the hair too short, decrease the angle of the razor cutting into the hair.

B Match the side lengths to the top length by pulling the hair up to the original guide and cutting. Use horizontal partings. When the hair does not reach the guide, stop cutting. Once again, use a long stroke with the razor, to create texture on the opposite side.

C Match the back sections to the top by using horizontal partings. Direct the hair straight up starting at the top of the left back section. Then cut the top right section. Alternate cutting sections working down the back of the head until the hair no longer reaches the top length. Then resection the back into two sections. Cut the hair using horizontal partings to a hanging length of 4 inches. Work up the head using a 0 degree elevation until the hair no longer reaches the guide.

D Comb all the air forward toward the face. Cut the bangs from the bridge of the nose to the outside of the eyes. Then direct the hair at the sides at an angle of 45 degrees forward and down. Cut a straight line from the bang to the longest length at the side. If the hair falls too heavy at the sides, use the point of the razor to carve out the bulk and create a light fringe by picking up 1/2 inch thick strands of hair. Run the razor down the shaft of the strands beveling the last 2 inches of the strand.

9

Styling

E Use a blowdryer and vent brush to create this style. First remove most of the moisture from the hair while directing the hair towards the face. Start at the top front hairline. Dry the hair forward and under. As you work back from the hairline elevate the hair away from the scalp to create height on top.

F Dry the sides next. Start at the hairline. Completely dry one section the width of the brush before moving back. Direct the hair toward the face. Bend the hair under while drying. Work diagonally back toward the crown.

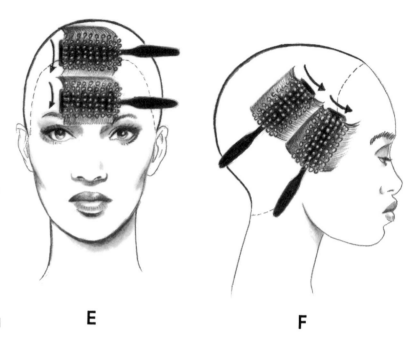

E   F

G Move to the back behind the ear. Follow the angle of the hairline and direct the hair under and forward toward the face. Move diagonally up to the center of the back section. This will leave the center of the hairline in the back and the crown to be dried next.

H First dry the bottom of the back down and under. Then move up to the crown and divide it into a left and right section. Dry each section of the crown down and forward. Next dry the very top of the crown up and back to create height.

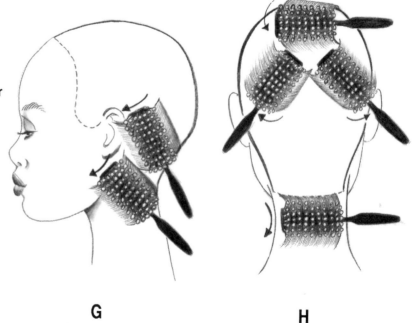

G   H

Cutting

A This cutting is for one length hair. Make a bang section approximately 1 inch deep from the outside of each eye. Cut the bang to just below the eyebrows. Lightly remove the weight from the bangs by laying the razor flat on the hair shaft and sliding down to the ends.

B Next divide the hair into three sections. Separate the top from the bottom at the crest line. Start at the temples and part off to the occipital. Divide the top down the middle into two sections from the center of the hairline to the occipital. Start at the top right side, take a vertical parting and direct the hair forward. Lay the razor flat on the strand and use medium pressure, remove varied lengths from 2 inches out from the scalp to 2 inches from the ends. If this is too much hair to handle, subdivide the parting and work from the top of the parting to the bottom. Follow this procedure throughout this section. Repeat on the left side.

C Once again, use vertical partings below the crest line. Start the cut at the hairline. The same procedure will be used on the sides as on the top until reaching the area behind the ear. In the back, direct the hair down to cut. The razor will still be held flat to the hair shaft. Start the cut 2 inches out from the scalp and stop 2 inches from the end.

D Take a section 1 inch deep around the entire hairline. Comb this hair forward. Starting at the corner of the bang, slide the razor down the strands and out from the head gradually increasing the length of the hair to form a "C" cut. This shorter hair underneath the longer texturized hair will support the style.

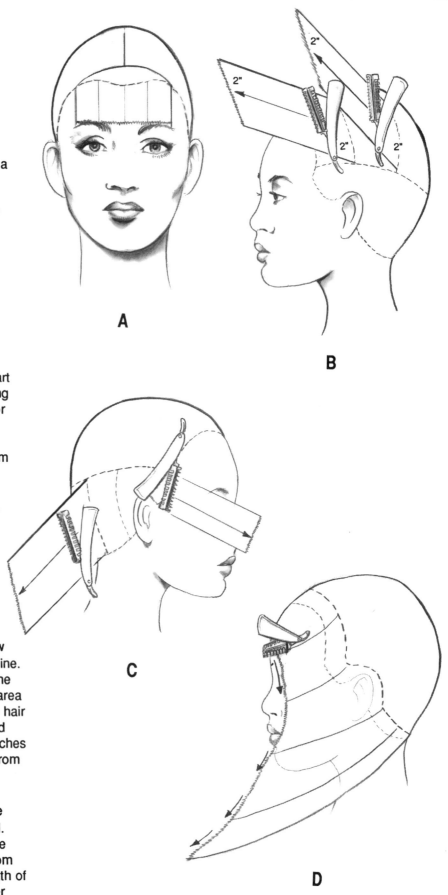

A

B

C

D

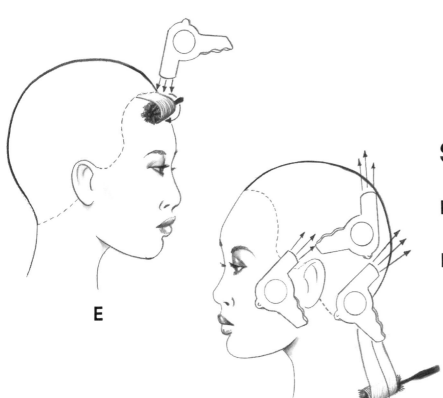

E

F

Styling

E Dry the bangs under and forward around a large round brush.

F Next dry all the hair using your fingers to direct the hair diagonally up and back on the sides. In the back you will start at the top and lift the hair up drying from the scalp to the ends. Use your fingers to tousle and separate the hair as you dry. Dry the nape down and around a large round brush.

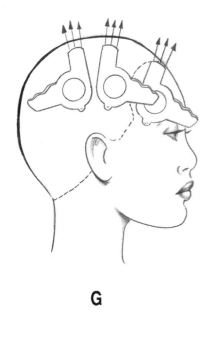

G

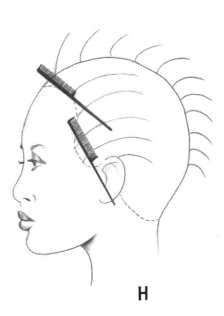

H

G Dry the top up from the head towards the back. Continue to use your fingers to create direction and separation at the top. Start at the back of the top section and work to the front.

H Backcomb the hair 4 inches up the strand on top. Gradually decrease the depth of the backcombing as you work down the sides and the back. The nape will not need to be backcombed. Lightly brush the hair back away from the face.

Cut

5

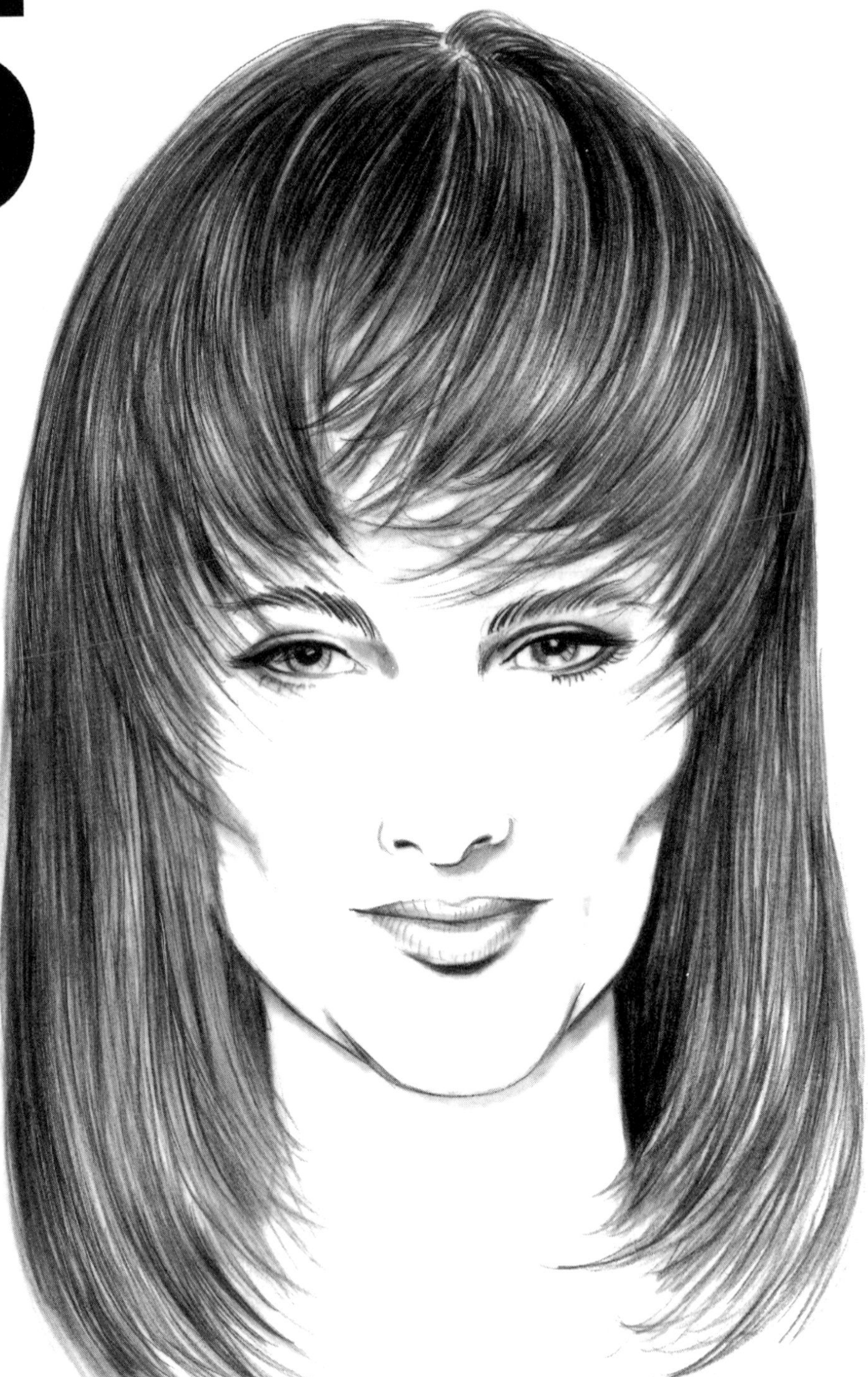

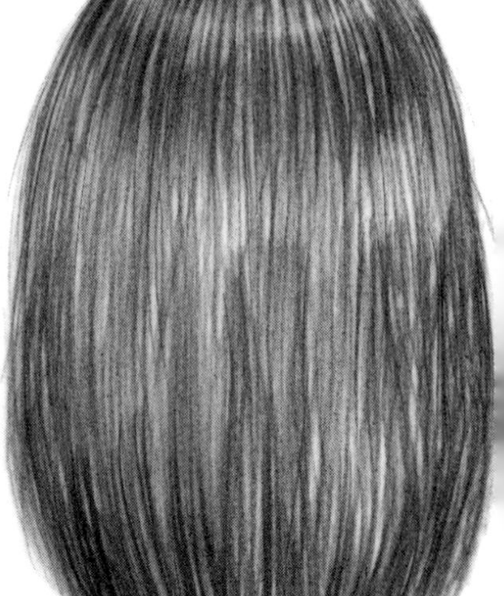

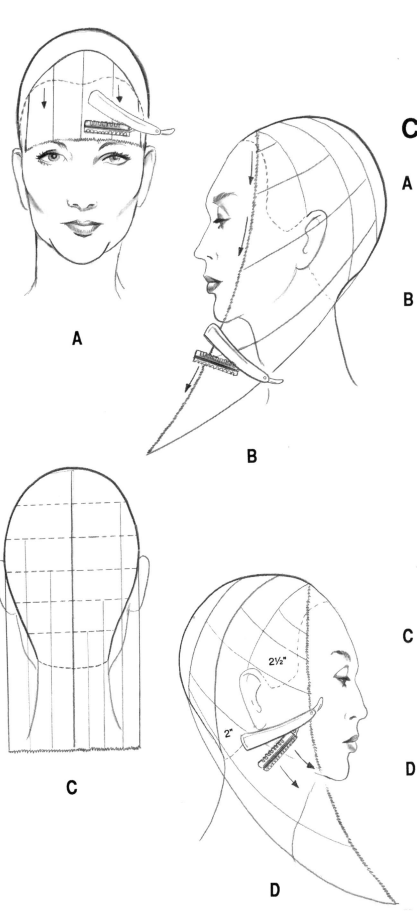

Cutting

A Make a horseshoe bang section 3 inches deep. Use a long stroke with the razor to remove length and bulk as you cut the length. Cut the bangs in the curve, shorter at the bridge of the nose to longer at the corner of the eyes.

B Divide the head into two sections from the center of the front hairline to the center of the nape. Use curved diagonal partings from the center part at the top of the head to the lower hairline. Direct the hair forward. Holding the razor perpendicular to the hair cut from the outside bang length to the longest hair in the parting. Continue to work to the back pulling the hair forward until the hair no longer reaches the guide. Repeat on the opposite side.

C Resection the back into two sections from the center of the top of the crown to center of the hairline at the nape. Cut the hair parallel to the floor using horizontal partings and a short stroke with the razor. Work up the head until the hair no longer reaches the guide.

D Make a curved parting, starting at the intersection of the bang and side section, to 1 inch behind the ear and down the nape. Starting at the top of this parting, lay the razor flat on the strand and gently remove some of the weight from the hair. Work down the section until all the hair in this section has been texturized. Repeat on the opposite side.

Styling

E Dry the bangs forward and under using a vent brush.

F Dry the sides forward toward the face and under using a vent brush.

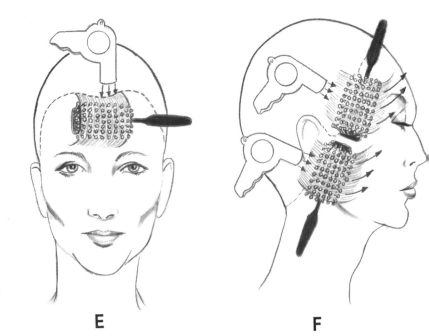

E

F

G Dry the back under using a large round brush. Start at the bottom and work up to the top. At the top of the crown, overdirect the hair on base and dry to create volume.

H Put a small amount of gel on your finger tips. Then pinch small strands of hair together by holding the hair 2 to 3 inches up the strand and then sliding down to the end while directing the hair to the face. Continue doing this around the front hairline and down the sides.

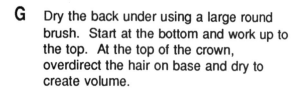

G

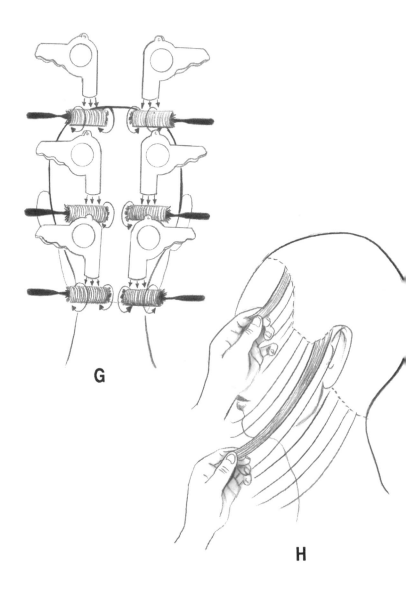

H

Cutting

A Divide the head into four cutting sections from the center of the forehead to the center of the hairline at the nape and from ear to ear over the high point of the crown. Establish your guide by making a 1-inch parting at the hairline in the nape. Make another 1-inch parting at the bottom of the side sections above the ear. Keep the head in a straight upright position. Cut the hair at 0 degree elevation to a hanging length of 5 inches. Match the side length to the back length and cut horizontal to the floor. Repeat on the opposite side. Check and make sure both sides are the same length.

NOTE: Use a very short stroke when cutting a bob with the razor.

B After establishing your guide, return to the back sections. Use horizontal partings to cut the back section. Start at the bottom of the left section. Work up the back alternating from the left to the right section. Comb the first parting straight down at 0 degree elevation to the guide. Cut the hair to the same length as the guide. Be sure to cut parallel to the floor to create a square back.

C Use horizontal partings on the sides. Start at the bottom of the section and work up to the top. Use 0 degree elevation and keep the head in an upright position. Match the guide length and cut parallel to the floor. After completing one side, repeat the same process on the other side.

D After completing the cut, resection into the same four sections. Make two horizontal partings in the back. Lay the razor flat against the strands and using light pressure taper the last 3 inches of the strand out to the ends. Be sure to taper the underside of the hair. This will enable the hair to turn under. Start at the bottom of each section and work up to the top of all four sections.

NOTE: Be sure to use very light pressure or you will remove too much hair.

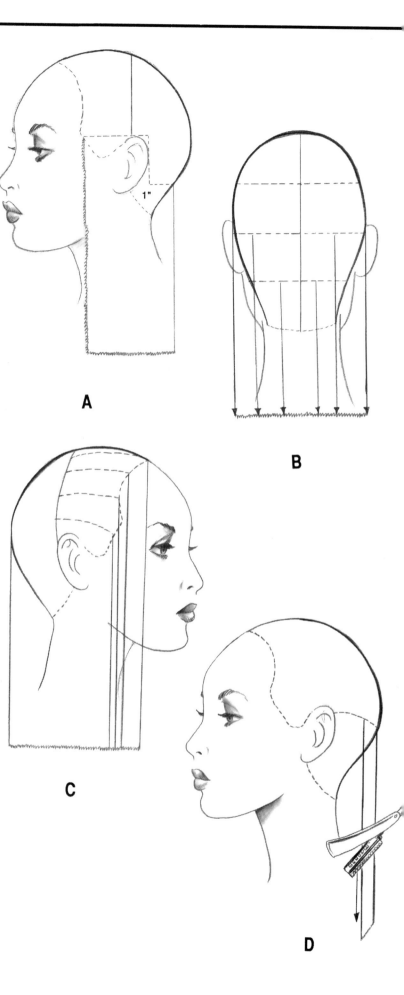

A

B

C

D

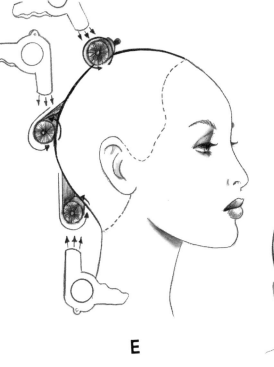

E

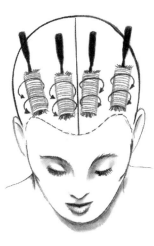

F

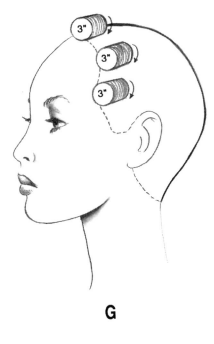

G

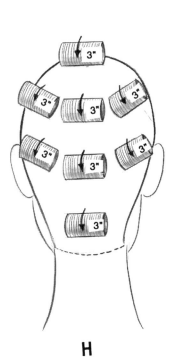

H

Styling

E Start drying the hair in the back. Use a blowdryer and a large round brush to create volume. Start at the nape and work up to the top. Your partings should be the same diameter as the round brush. Roll the hair around the brush to go under. At the bottom sections of the back, the hair will be dried off base, the center sections 1/2 on base, and the top on base.

F Dry the hair in the front parallel to the center part. Start on the side and work up. The sides and top will be rolled to go under. Dry bottom of the sides 1/2 on base and the top on base.

G Set the hair on 3-inch diameter velcro rollers. Spray the dry hair lightly with a sculpting spray from the base to the ends. Start at the center part at the front hairline. Roll the hair on base and under. The size of the parting rolled should be the same as the size of the roller. Work down the sides placing the second roller 1/2 on base. If there is room for a third roller place it off base.

H Start setting the hair at the top of the crown. Set your first roller on base and back. Roll the hair under and back to the nape. The second roller should be on base, and the third roller 1/2 on base, and the fourth roller off base. Set the side sections of the back straight down starting on base at the top, 1/2 base in the center of the section and off base at the bottom. Set the head under a hot dryer for 15 minutes. Allow the hair to cool before removing the rollers.

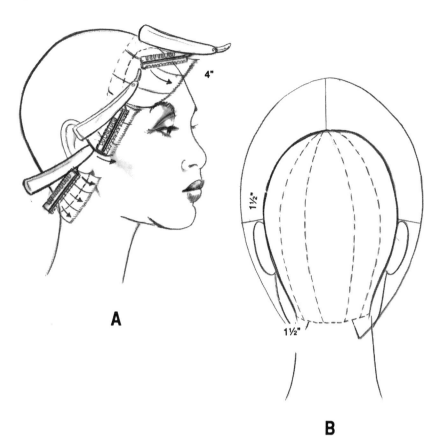

A

B

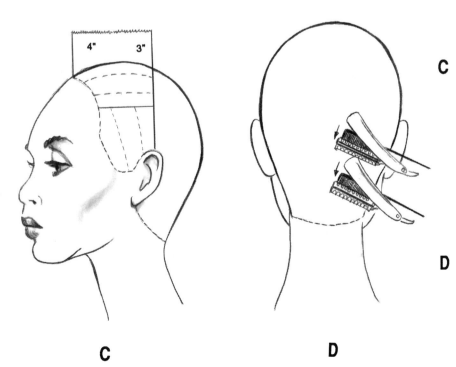

C

D

Cutting

A   Make a section around the perimeter of the entire hairline 1 inch deep. Lay the razor at a 30 degree angle to the hair and remove length and bulk at the same time. Start at the front and work to the back, then bang area will be 4 inches in length. Reduce the length at the temporal area to 1 1/2 inches and blend into the bang section. The remaining length of the rest of the perimeter will be 1 1/2 inches.

B   Start the interior cut at the top of the crown. Use a 1-inch stroke with medium pressure to cut the remainder of this style. Use pie shape partings from the center to the crown to cut the back. The length at the top of the crown will be 3 inches. Gradually decrease the length as you work down the parting reducing the length to 1 1/2 inches at the occipital area. Cut the remainder of the parting to match the exterior guide length of 1 1/2 inches.

C   The front half of the head will be divided into three sections. A center top section and the two remaining sides. Partings will be from the bang to the crown of the top section. Match the length at the crown to the length at the bang and remove the excess length. Complete the rest of the top section holding the hair at 90 degrees from the head. Match the sides to the top length and blend into the perimeter length of 1 1/2 inches using vertical partings.

D   Texturize the nape by using the razor over comb rotation method. Comb the hair down and follow it with the razor held flat to the head with no pressure. Lift the comb after a 1 1/2 inch stroke and then lift the razor at the same spot. Work down to the hairline.

Styling

E   Dry the perimeter at the sides and in the back around a small diameter round brush. The hair is dried under and forward at the temples, the sides over the ears, and the sides of the back hairline. The center back hairline will be dried under.

F   Dry side sections holding a round brush diagonally up and towards the face. Work from the front of the section to the back. Dry the top diagonally towards the right side. Start at bang area and work back to the crown.

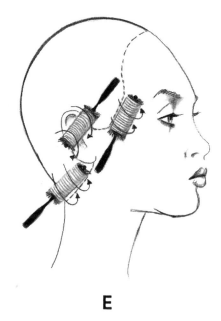

E

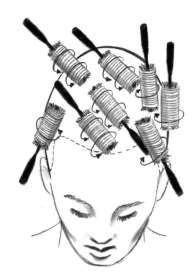

F

G   Dry the back in a swirl pattern. Make the swirl of center and to the left side of the center of the crown. Everything above the swirl will be diagonally up towards the front. Everything below the swirl will be dried diagonally down to the sides of the back.

H   All the hair will be brushed starting at the swirl and moving to the hairline. All the hair above the swirl will be directed to the sides and over the top. Below the swirl, the hair will be directed across the back diagonally to the side hairline. Brush the hair across the top from the left to the right side, curving toward the hairline at the right side.

G

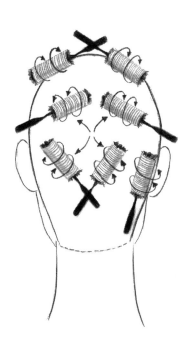

H

Cutting

A Hair must be chemically relaxed. Make a part from the top of each ear to the occipital bone. Start your cut in the middle of this section. Comb the hair straight up to the part line. Cut the hair horizontally 1 inch out from the part. Use a firm short stroke to cut the hair. Follow this procedure on the left and right sides of the nape. Then comb the nape section down and cut to a hanging length of 2 inches at the center of the hairline.

B The crown section will be from ear to ear over the top of the head. Use horizontal partings in the crown section. Start at the bottom of the section and work up. Match to the length of the previously cut section. The partings will be held parallel to the floor. The length of the hair will gradually increase as you work to the top of the crown.

C Divide the front into three sections, the top and two sides. Start at the bottom of the side section. Use horizontal partings to cut the sides. Blend the first parting into the back 1-inch length. Work up the side sections holding the hair straight out from the head. Match to the length in the back and the previously cut hair.

D Make a parting down the center of the top section from the hairline to the crown. Cut the top section to a length of 3 1/2 inches. Continue using partings from the hairline to the crown to cut the top section. Direct the hair straight up to cut. When the top is completed, comb the hair in the top section forward and cut to a length of 3 1/2 inches.

NOTE: If the hair is very thick or coarse, remove excess weight by laying the razor flat to the strand and sliding out to the ends using light pressure.

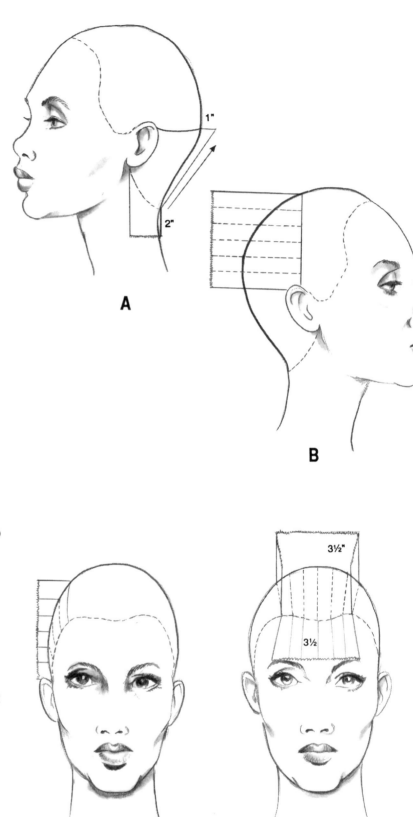

A

B

C

D

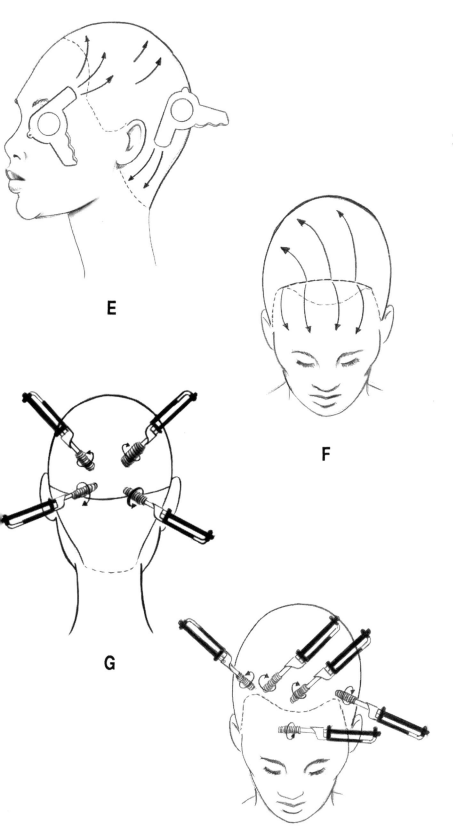

E

F

G

H

Styling

E Start in the nape and dry the hair down and under. Next, begin in the center of the crown and dry the hair diagonally up and back. Work toward the side sections continuing to direct the hair up and back.

F Dry a small section at the hairline down to the forehead using a vent brush. Dry the top section back and diagonally across the head from left to right.

G Use a large barrel curling iron to set the style. Start in crown section and curl the bottom of the crown diagonally to the center and under. Curl the center of the crown diagonally up and to the center. Curl the top of the crown straight back. The nape will not require any curl.

H Curl the sides diagonally up and back. The top will be curled diagonally back and across the top from left to right. Curl the last 1 inch of the bangs.

Cutting

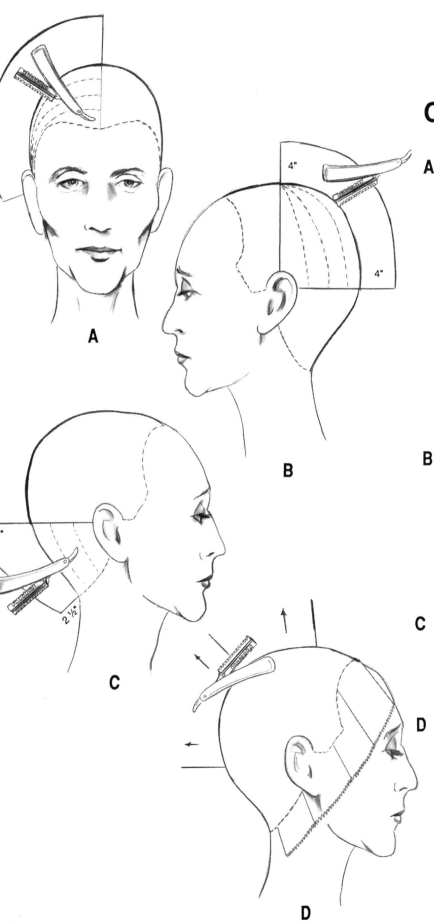

A

B

C

D

A Section the hair into four sections. Divide the nape from the crown, from the top of the ear to the top of the other ear, and across the occipital bone. Divide the crown across the top from ear to ear. Make a center part to the crown, making the two front sections. Using vertical partings begin cutting at the back of the right front section. Hold the hair at 90 degrees from the head and cut to a length of 4 inches. Cut from the top of the parting to the bottom. Use a long stroke with the razor to remove length and to taper the ends. Match the second parting to the first and continue to cut to a 4-inch length. When the right side is completed, follow the same procedure on the left side.

B Cut the crown section next. The crown will be an equal blend cut to a length of 4 inches. Use vertical partings and start on the right side of the section. Cut from the top of the parting to the bottom and work from right to left. Remember to hold the hair at 90 degrees from the head.

C Cut the nape section to a hanging length of 2 1/2 inches. Use vertical partings to blend from the 4-inch length at the crown to the shorter length at the perimeter of the nape.

D Clean up the perimeter line by combing it into its natural fall and evening the ends. Then beginning in the front and working to the back lightly texture the last 3 inches of the strands to allow for better backcombing. Texturize by holding a parting approximately 1 inch wide by 3 inches across, straight out from the head, then lay the razor flat on the shaft and gently slide the razor out to the ends.

Styling

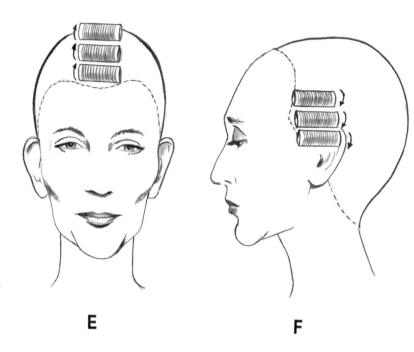

E   Use a firm hold styling lotion to set the hair. Use medium diameter rollers on top. Set the hair straight back and on base.

F   Set the sides using the same rollers as the top. The hair will be rolled down. The first roller will be on base to help blend the style with the top. Second roller will be 1/2 on base and the third roller off base.

E   F

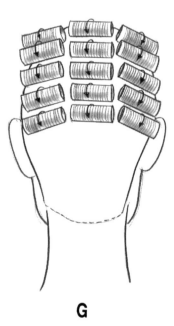

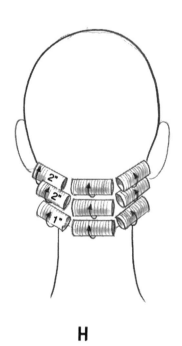

G   Roll the crown using the same size rollers as the top and sides. Roll them under and down. Divide the crown into three vertical panels. The first three rollers will be set on base. The last two in each panel will be set 1/2 on base.

H   Roll the nape section up. Divide the nape section into the same three panels as the crown. Start at the top of the panel, use a roller one size smaller than the rollers used in the crown. Use smaller rollers as you work down to the hairline.

G   H

10

Cutting

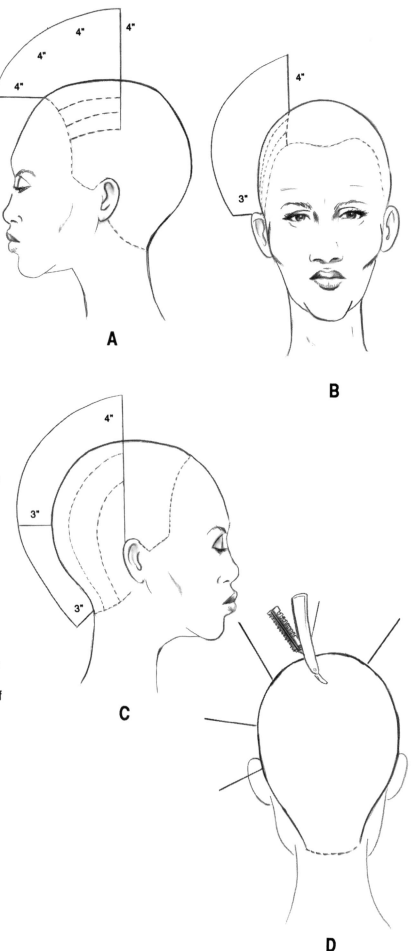

A This style requires the hair to be chemically relaxed. Divide the head into five sections. Three sections in the front, a top and two side sections, and two in the back, a left and right section. Use a center parting from the back of the top section to the front. Cut the hair at 90 degrees from the head to a length of 4 inches. Match the length of the remaining hair to the guide by directing cut parting to the parting to be cut. The partings should be 1 inch wide. Repeat on both sides of the top section.

B Use vertical partings to cut the sides. Start at the back of the section. Work from the top of the parting down to the bottom of the parting. Gradually decrease the length of the hair from 4 inches at the top of the side section to 3 inches to the bottom.

C Use pie shape partings to cut the back. Create a guide down the middle of the back section from the center of the crown to the center of the hairline at the nape. Cut from a length of 4 inches at the top of the crown to a length of 3 inches at the occipital. Maintain a 3-inch length throughout the nape. Work from the center to the right side section and then the left back section.

D Hold the hair out 90 degrees from the head and remove 30 percent of the bulk starting 1 inch from the scalp to 1/2 inch from the ends. Keep the angle of the razor flat on the hair and use light pressure.

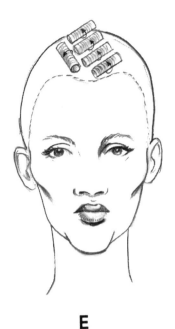

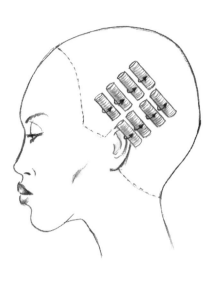

E

F

Styling

E   Set the hair using a medium to strong hold sculpting lotion or gel. Begin on top at the hairline. Set a row of 1-inch rollers on base diagonally back and to the right. Set one roller perpendicular to the diagonal row and forward. This roller will be 1/2 on base.

F   Set the left side using two rows of roller set diagonally up and back. These top rows will be set on base using 3/4 inch rollers and the bottom row will be set on base using 1/2 rollers.

NOTE: If there is not enough room for two full length rollers on the left side, use filler or 3/4 length rollers on the bottom row.

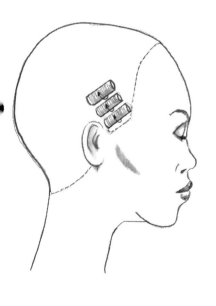

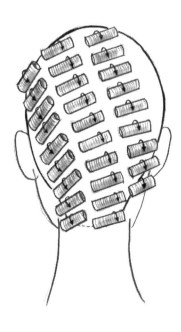

G

H

G   The right side will hold only one row of rollers. Set a row of diagonal rollers up and back. The first two rollers will be 1/2-inch rollers and the third roller will be a 3/4-inch roller.

H   The diagonal row of 1-inch rollers on top will continue to the right side of the back. Switch to 3/4-inch rollers after crossing the high point of the crown. The top diagonal row on the right side will continue into the back down to the center. Continue to use 3/4 inch rollers. Use 3/4-inch rollers to complete the left side of the back. After drying, smooth the entire head following the same pattern as the set.

11

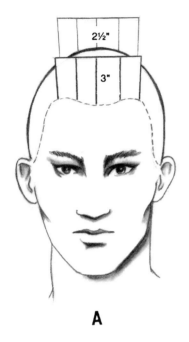

A

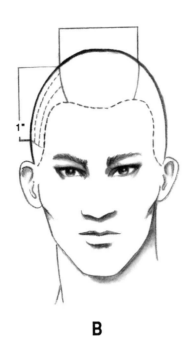

B

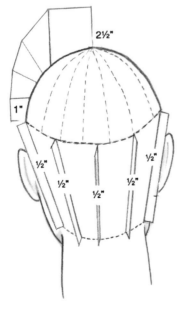

C

D

Cutting

A The head will be divided into four sections, below the occipital bone, the crown, and the two side sections. Cut the top into a square shape. Start at the highest point of the top section. Direct the hair straight up and cut parallel to the floor. Cut to a length of 2 1/2 inches. Use horizontal partings as you work forward to the hairline. Direct the previously cut parting to the parting to be cut. This will be your guide. The length will increase slightly as you work toward the hairline due to the curvature of the head. The top should form a plane which is parallel to the floor.

B To cut the side sections use the top length as your guide. Direct the hair out from the sides parallel to the floor. Cut the hair perpendicular to the floor. Start in the back of the side section and work to the hairline. This will create a box or square form. Cut the hair as blunt as possible with the razor.

C Cut the crown section by using the top and side sections as your guide. Start with a 2 1/2 inch length at the top and work down to a 1-inch length. Use pie shape parting to work around the head. Work from the front hairline through the crown directing the hair back and following with the razor. When the comb can move freely through the hair without any resistance, the hair is tapered correctly.

D Use a razor over comb rotation to remove the weight and taper the ends above the crestline. The area below the crestline should be cut to a 1/2 inch or shorter length.

Styling

E Dry the area below the crestline first. Direct the hair in the nape down and towards the hairline. The sides below the crestline will be dried diagonally back and down.

F Dry the center of the crown straight down around small diameter round brush. The side of the crown will be directed diagonally to the center of the back.

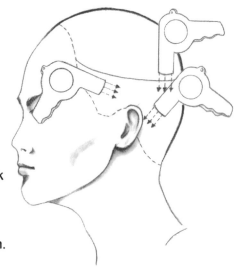

E

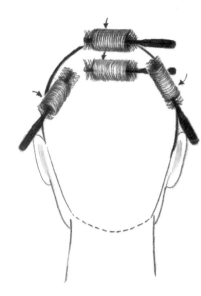

F

G Dry the top straight back. Start at the back of the section and work to the front. All the hair in the top section should be dried on base to accent the square form.

H The side will be dried diagonally up and to the back. Start at the back of the side section and work to the hairline.

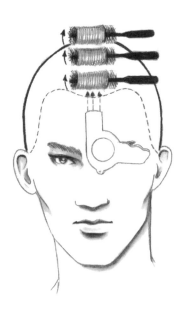

G

H

Cutting

A Divide the back into a crown section and a nape section. Using vertical partings, cut the nape from a length of 1 inch at the occipital to a length of 3 inches at the hairline. Slide the razor down the parting from the occipital to the hairline increasing the length as you move down. Use the top length at the nape as a guide to start cutting the crown. Hold the guide straight out from the head, drop the first horizontal parting down to the guide. Use a long tapering stroke with the razor to cut the entire crown. Continue to drop the hair to the previously cut hair elevating slightly as you work up the head.

B Take a small sideburn section in front of the ears and cut to a length of 4 inches. Lightly taper the sideburns by sliding the razor down the strand using little pressure. Cut the heavy side of the front using horizontal partings and no elevation. Continue using a long tapering stroke with the razor. Direct the bang area of the heavy side back to the hairline in front of the ear to increase the length in front. Cut parallel to the floor. Blend into the back using horizontal partings.

C Cut the light side at 0 degree elevation around the ear. Use horizontal partings and long tapering strokes with the razor. Blend into the back length using horizontal partings.

D Section 1 inch deep around the entire front hairline. Start on the left side. Cut from the left hairline to up over the left eye. At the center of the left eye overdirect the remaining hair over to the left side. Blend from the short length over the eye to the longest length over the right eye.

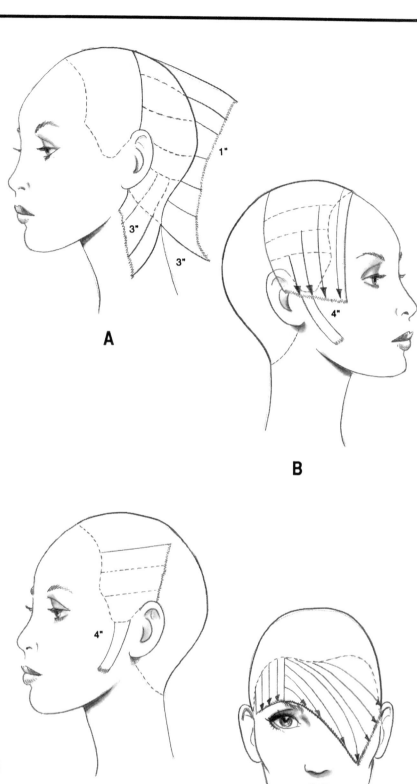

Styling

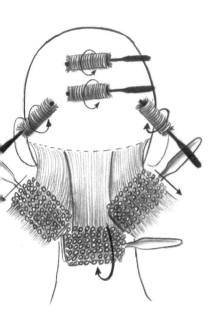

E

F

E Dry the nape area flat to the head using a vent brush. Turn the hairline under at the nape. Dry the lower portion of the crown using a round brush. Direct the hair under and straight down in the center of the crown. The sides of the crown will be directed under and diagonally to the center.

F The top of the crown will be dried to create volume. Place a large round brush on base at the top of the crown (highest point on top of the head), dry the hair and let cool before removing the brush. Repeat this process on each side of the crown holding the brush vertical and directing the hair back.

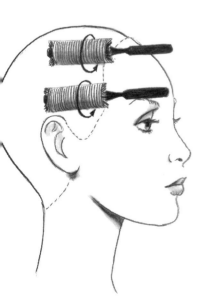

G

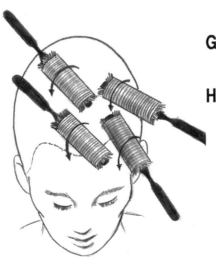

H

G Dry the heavy side of the front under using a large diameter round brush.

H Dry the light side of the front forward using a round brush. Start at the hairline and work back to the crown. Do not create height on the sides. The top section will be dried diagonally forward and under. Dry the back of the top off base and the front of the top section on base to create height at the bang. Hold the brush off base to dry the heavy side. Backcomb the crown and the bang section to accent the height and fullness.

Cut
13

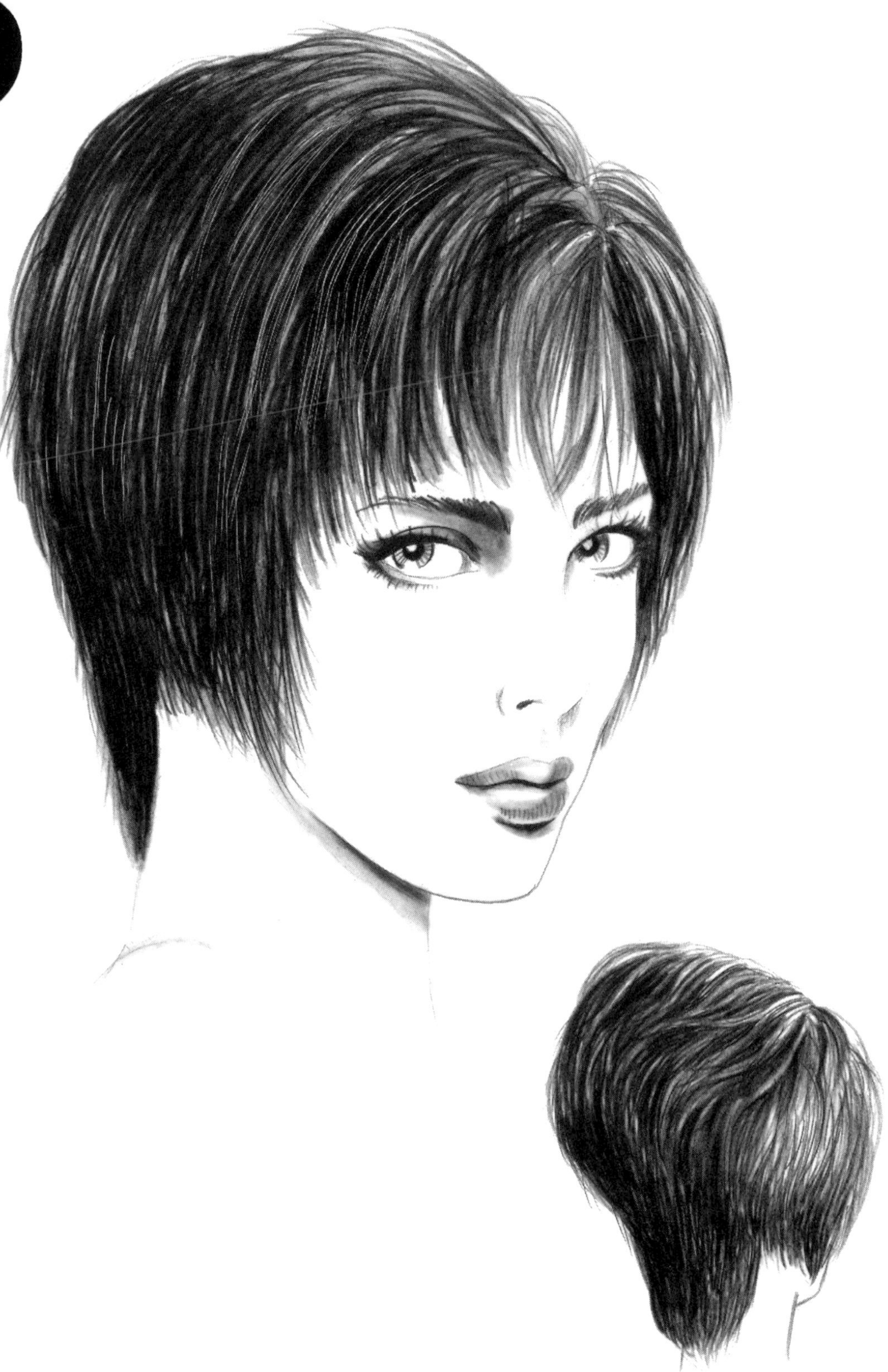

Cutting

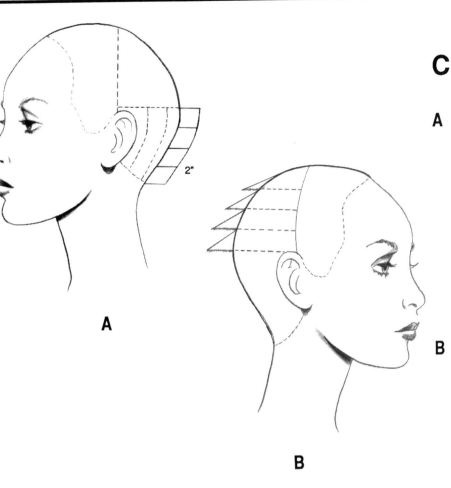

A

B

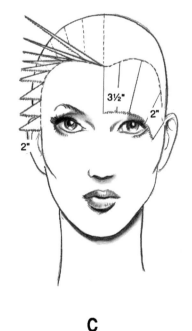

C

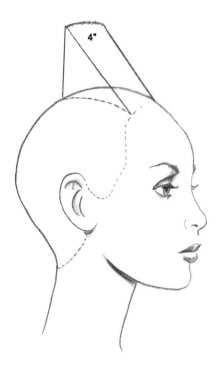

D

A   Divide the head into four sections. Separate the nape from the crown at the top of the ears to the occipital bone. Divide the crown from the front at the back of the ears to the high point of the top of the head. Make a center part to divide the front into two sections. Begin the cut in the nape. Make a vertical guide down the center of the nape. Cut to a length of 2 inches using a 1-inch stroke with the razor while holding the hair at a 90 degree angle from the head. Cut the remaining hair in the nape 2 inches in length.

B   Use a horizontal parting across the nape as the guide for the crown. Hold the guide at a 90 degree angle from the head. Make a horizontal parting across the bottom of the crown. Direct the hair down to the guide and cut. Continue to use a 1-inch stroke. Start at the center of the crown and work to each side. Hold the parting which has just been cut out to 90 degrees. Drop the next parting down to the first cut and match the length. Continue to work up the crown holding the cut hair at 90 degrees and dropping the hair to be cut down to the guide. This will gradually increase the length as you move up the head.

C   Start at the bottom of the side section. Use horizontal partings to cut the sides. Hold the hair straight out at the hairline and direct the hair from the part line down. This will gradually increase the length of the hair in the parting. Cut to a 2-inch length at the hairline using a 1-inch stroke with the razor. Continue working up the sides using the same method as the crown section. After completing both sides, direct all the hair forward and cut a 3 1/2 inch bang blending into a 2-inch length at the temple.

D   Comb all the hair on top, back to the top of the crown. Cut to a length of 4 inches at the stationary guide. Comb all the hair in the crown up to the top of the crown and cut to the same 4-inch length.

Styling

E Dry the nape down and flat to the head. Brush the hair down and follow with the blowdryer.

F Dry the crown using a large round brush (2-inch diameter). Overdirect the hair in the top of the crown up and back to develop height. As you work down the crown decrease the amount of over direction. At the bottom of the crown do not overdirect the hair so that it will blend into the nape.

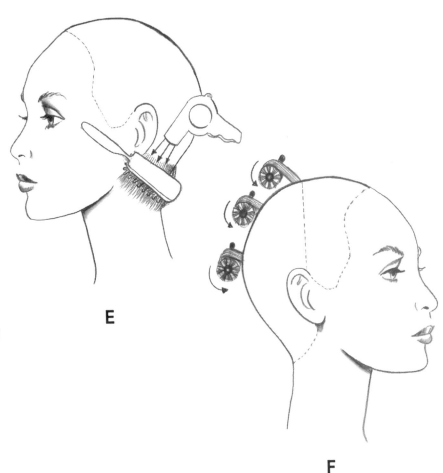

E

F

G The hairline of the front section is dried forward using a vent brush. Direct all the hair at the sides forward. Make a part on the left side. The heavy side of the part will be dried over the top of the head with no volume. The light side at the part will be dried down again with no volume.

H Set the crown section on 2-inch velcro rollers. Spray each section with a sculpting spray before setting. The top roller in the crown will be on base. The side roller in the crown will be 1/2 base and the bottom row of rollers will be off base. Place under a hot dryer for 15 minutes, allow to cool and then style.

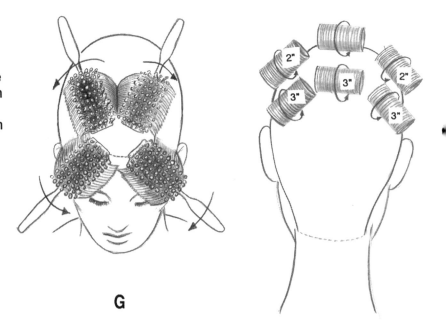

G

H

Cutting

A Section the hair into five sections, two in the back, one in front, and the two sides. Cut the top front section to a length of 3 inches. Be sure to follow the curvature of the head as you work from the back of the top section to the front of the top section. Cut with a quick short stroke of the razor.

B Cut the sides using the top section as a guide. Decrease the length as you work down the side from 3 1/2 inches to 1 inch over the ear. Work from the back of the side sections to the hairline.

C Cut the back using the top section as your guide. Start in the middle of the back using vertical partings. Cut from the top to the bottom of the nape. Gradually reduce the length as you work down. Cut the top right side first, then cut from the center to the left side. After all the interior layering is complete cut your perimeter lengths. The hairline should be cut to a length of 1 inch from the nape to the sideburn. Comb the bangs down and cut to a length of 3 inches. Then blend the bangs into the sides.

D Remove bulk throughout the entire head using razor over comb technique. Start on the left side and work back to the nape on the left side. Repeat this procedure on the right side ending in the right nape. Then using the razor over comb method, remove the bulk on top working from the front hairline through the center of the crown and down to the nape.

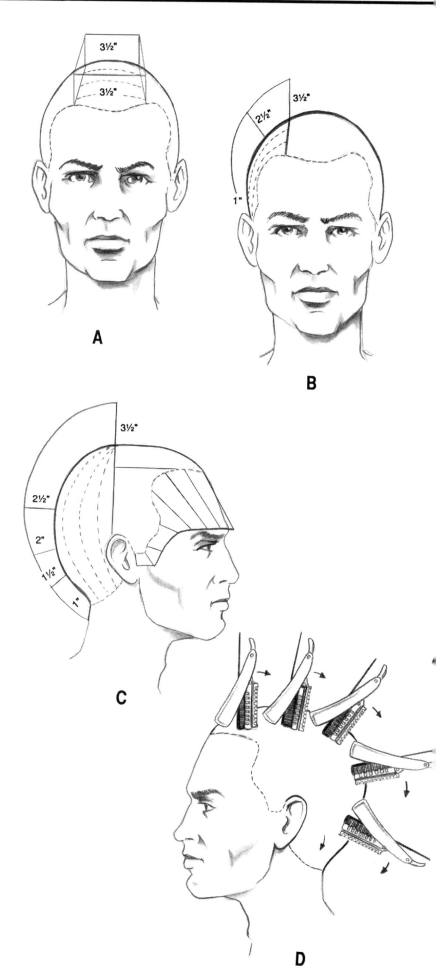

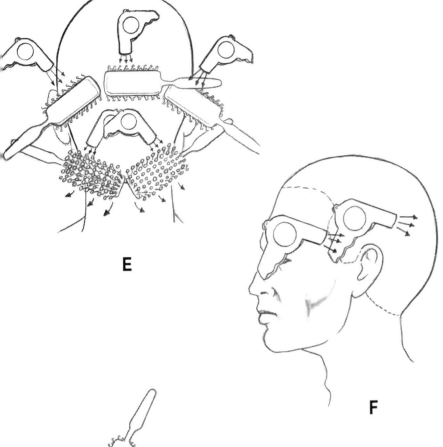

E

F

Styling

E Begin drying the hair by directing the perimeter of the nape down and under. The center of the crown is directed down. Direct the sides of the crown diagonally down and to the center. Use a vent brush to dry the hair. When giving the hair direction, the bristles of the brush should be placed down into the hair. The brush is moved through the hair in the desired direction with the blowdryer following behind.

F Dry the light side of the part back to the crown. Work from the back of the head to the front hairline.

G Dry the center of the top first. Dry the hair from the part across the top. Direct the bang area diagonally back and away from the part.

H Direct the heavy side of the part back to the crown.

G

H

Cutting

A Take a curved section at the side hairline, from the eyebrow to just in front of the ear and direct the hair forward cutting a C shape. Cut from the top of the section down holding the razor perpendicular to the strand. Slowly slide the razor down the strand gradually increasing the length.

B Cut the hair on top holding it straight up to form a square shape. Start at the back of the crown cutting the hair 2 inches in length. Work toward the front using horizontal partings. The length will gradually increase as you move to the front due to the curvature of the head. Use a long stroke with the razor to create an airy look on the ends and to give support to the style. Cut the hair perpendicular to the floor using vertical partings. Start at the top of the side section and cut down the parting. Use the length on top as your guide. The length of the hair will decrease as you work down the head due to the curvature of the head. Be sure not to recut the sideburn section. Start at the back of the section and work forward.

C Cut the nape section using vertical partings from a length of 2 inches at the top to 1 inch at the hairline.

D Cut the crown to a length of 2 inches. Make a center guide down the middle of the crown. Match the length at the top to the length at the top of the nape. After completing the nape use horizontal partings from the side sections into the crown section. Direct the hair to 90 degrees and blend from the front to the back.

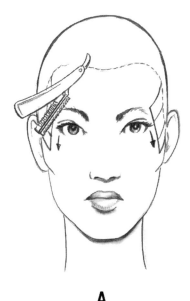

A

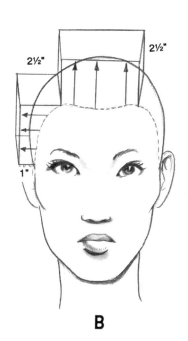

B

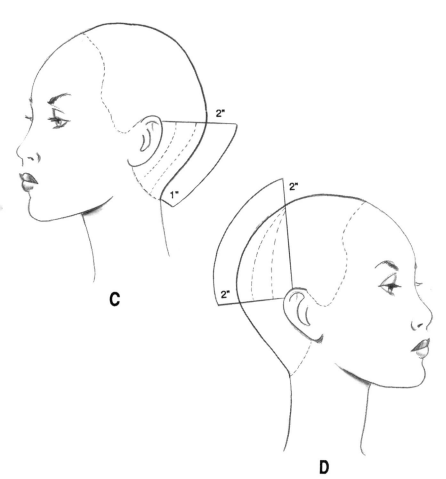

C

D

Styling

E Dry each side of the nape section toward the center. Continue drying the hair toward the back as you work into the side sections below the crest line. Dry the sideburns toward the face.

F Prepare the hair using a strong hold gel or sculpting spray. Dry all the hair in the top setting straight up at 90 degrees. Spray small amounts of sculpting or hair spray as you dry to give added support to the style.

G Dry the sides above the crest line up and slightly diagonal towards the back. Spray with sculpting spray as you dry the hair to hold it up.

H Dry the center of the crown straight up. Dry the sides of the crown up and toward the center. Continue spraying with sculpting spray as you dry. After drying all the hair, look for areas that are not uniform. Spray the area with sculpting spray and lift with a pic comb.

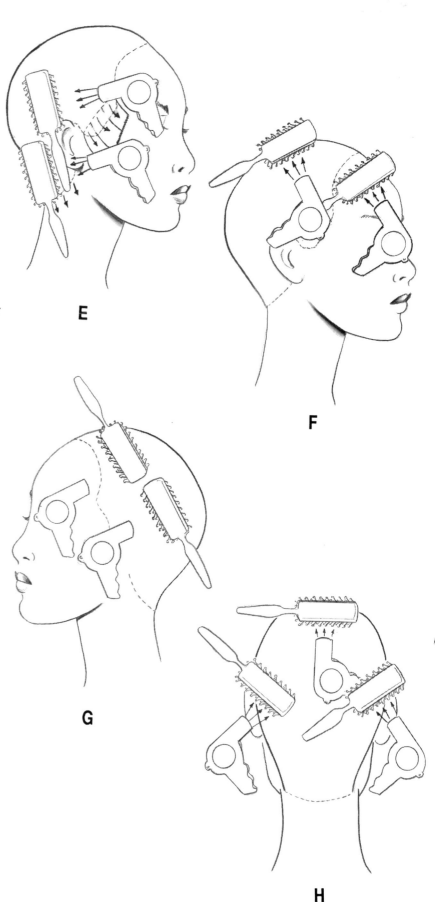

E

F

G

H

Grading Record

Note: Each time a style is performed, your instructor will grade it on a scale of 1 to 10. A grade of 5 to 6 is average, and any grade below that shows a serious need for improvement.

Cut 1

Date	Grade	Instructor	Instructor's Comments

Cut 2

Date	Grade	Instructor	Instructor's Comments

Cut 3

Date	Grade	Instructor	Instructor's Comments

Cut 4

Date	Grade	Instructor	Instructor's Comments

Cut 5

Date	Grade	Instructor	Instructor's Comments

Cut 6

Date	Grade	Instructor	Instructor's Comments

Cut 7

Date	Grade	Instructor	Instructor's Comments

Cut 8

Date	Grade	Instructor	Instructor's Comments

Cut 9

Date	Grade	Instructor	Instructor's Comments

Cut 10

Date	Grade	Instructor	Instructor's Comments

Cut 11

Date	Grade	Instructor	Instructor's Comments

Cut 12

Date	Grade	Instructor	Instructor's Comments

Cut 13

Date	Grade	Instructor	Instructor's Comments

Cut 14

Date	Grade	Instructor	Instructor's Comments

Cut 15

Date	Grade	Instructor	Instructor's Comments

NOTES

NOTES

NOTES

NOTES

NOTES